# SOIL AND FERTILITY

Building the Living Ground Everything Else Depends On

**Hudson Briggs**

The Homestead Starter Series — Book 17

## Soil Testing, Safety, and Fertility Disclaimer

This book provides general educational information about understanding, testing, and improving garden soil at the home and small homestead scale. It is not professional agricultural, environmental, legal, medical, food-safety, laboratory, pest-control, engineering, or financial advice.

Soils vary by region, site history, texture, drainage, contamination risk, and management. Amendment rates, costs, ratios, timelines, and results are illustrative ranges, not guarantees. Base lime, sulfur, fertilizer, and other amendment rates on a laboratory soil test for your own site. Test for lead before growing food in older urban lots, near old painted structures, or on land with unknown history. Know the source of imported manure, hay, straw, and compost because persistent herbicide residues can damage crops for multiple seasons. Follow product labels, local rules, and extension guidance, avoid over-applying manure or amendments, and treat soil building as a multi-year process.

## A Note Before You Begin

This book is a practical guide to understanding, testing, and improving garden soil at the home and small homestead scale. It is written for the grower who has been following recipes - add this, sprinkle that - and wants to understand what is actually happening underfoot.

Every rate, ratio, and cost figure here is an illustrative range drawn from widely accepted practice. Soils vary enormously, and the single most valuable thing you can do is have your own soil tested by your local extension service. A ten-dollar test beats every general figure in this or any other book.

Two cautions belong at the front. If you garden in an older urban lot, near an old painted structure, or on ground with unknown history, have the soil tested for lead before growing food in it. And never bring in manure, hay, or compost of unknown origin without knowing its source, because persistent herbicide residues can pass through and devastate a garden for years.

Finally, the honest frame this book returns to throughout: soil building is a multi-year project. No single amendment fixes soil. What fixes soil is a sustained practice, repeated season after season, and the good news is that it compounds.

# Introduction

## What You'll Learn

This introduction explains why soil is the foundation beneath every other book in this series, what "good soil" actually means in measurable terms, and how this book moves from understanding to testing to a long-term building program.

## The Problem Underneath Everything

Most gardening problems are soil problems wearing a costume.

The tomatoes that never sized up, the greens that turned bitter and bolted, the carrots that forked, the beds that flood in spring and crack in August, the plants that sit motionless for weeks after transplanting - these get blamed on weather, on seed, on pests, on luck. Very often they trace back to compacted, depleted, badly drained, or wrongly balanced ground.

This is also why so much garden advice fails. A grower reads that a crop needs fertilizer, applies fertilizer, and sees no improvement - because the nutrient was already present and simply unavailable at that soil's pH. Another grower doubles the compost and still has plants drowning, because the problem is compaction, not fertility. Without understanding the soil, you are guessing.

## Why It Matters

Soil is the only input in your garden that compounds. Seed is consumed each season. Fertilizer is used up. Water evaporates. But soil that you improve this year is better next year, and better still the year after - and it stays improved as long as you do not destroy it.

That makes soil the highest-leverage place to put your effort. A grower with excellent soil and ordinary seed will outproduce a grower with excellent seed and poor soil, every single year, with less watering, less fertilizer, and fewer problems.

## What "Good Soil" Actually Means

The phrase gets used loosely. In practical terms, good garden soil has four measurable qualities:

Structure - it holds together in crumbs, drains freely, and lets roots through without resistance. Organic matter - ideally around 5 percent, which feeds biology, holds water, and buffers everything else. A workable pH - for most vegetables, between 6.0 and 7.0, because pH controls whether nutrients are available at all. Adequate nutrients - enough of the major and minor elements, in a form plants can take up.

Those four are the outline of this book, and each is measurable, improvable, and worth understanding.

## How This Book Works

Chapters 1 through 3 build understanding: what soil is physically, what lives in it, and what your soil specifically is. Chapters 4 and 5 cover testing and pH - the diagnostic core, and the two chapters that save the most wasted money. Chapters 6 through 9 cover the building tools: organic matter, nutrients, mulch, and cover crops. Chapters 10 and 11 cover protecting soil from compaction and tillage damage, and diagnosing problems. Chapter 12 assembles a multi-year plan.

## Action Plan

Before you read further, order a soil test from your local extension service. It costs ten to thirty dollars, takes a few weeks to come back, and every chapter after this one is more useful when you have your own numbers in hand.

# Chapter 1: What Soil Actually Is

## What You'll Learn

This chapter covers the physical makeup of soil: the mineral fraction, organic matter, and - the part most growers overlook - the pore space that makes up roughly half of good soil.

## Why It Matters

Understanding soil physically explains why compaction is so destructive, why drainage and water-holding are two sides of the same property, and why "add more compost" is not the answer to every problem. It is the foundation for every decision later in the book.

## The Four Components

Good garden soil, by volume, is roughly:

**What Soil Is Made Of, By Volume**

| Component | Approximate Share | What It Does |
|---|---|---|
| Mineral particles | About 45 percent | Sand, silt, and clay; the physical skeleton |
| Organic matter | About 5 percent | Feeds biology, holds water and nutrients |
| Water | About 25 percent | Held in pore spaces, available to roots |
| Air | About 25 percent | Roots and soil life need oxygen |

Look at that table again and notice something most growers never think about: half of good soil is empty space. Water and air together account for about 50 percent of the volume, held in the pores between and within soil aggregates.

That single fact explains a great deal. It explains why compaction - which crushes pore space - is so damaging. It explains why roots suffocate in waterlogged soil (the water has displaced the air). And it explains why soil structure matters as much as soil content.

## Sand, Silt, and Clay

The mineral fraction is classified by particle size, and the proportions determine your soil's texture.

Sand particles are the largest. Sandy soil drains fast, warms early in spring, is easy to dig - and dries out quickly and holds few nutrients. It feels gritty.

Silt particles are medium. Silty soil holds water and nutrients better than sand and feels smooth or floury when dry.

Clay particles are the smallest, with an enormous surface area relative to volume. Clay soil holds water and nutrients extremely well - sometimes too well - drains slowly, compacts easily, and is heavy to work. It feels sticky when wet and sets hard when dry.

Loam is the balance point and the ideal: roughly 40 percent sand, 40 percent silt, and 20 percent clay. It drains adequately, holds water adequately, and works easily.

## Texture Cannot Be Changed - Structure Can

This distinction is one of the most useful in this book.

Texture - your ratio of sand to silt to clay - is essentially permanent. You are not going to change the mineral composition of a garden by adding material; the quantities required are absurd. Attempting to "fix" clay soil by adding sand is a classic and expensive mistake that, in the wrong proportions, can produce something closer to concrete.

Structure - how those particles clump together into crumbs and aggregates, and how much pore space exists between them - can be improved dramatically, and the tool for it is organic matter. This is why compost improves both sandy soil and clay soil, which sounds contradictory until you understand the mechanism: organic matter helps sandy soil hold water, and helps clay soil form aggregates that let water through.

So the practical rule: stop trying to change your texture, and put all your effort into structure.

## Common Mistakes

Adding sand to clay soil in an attempt to lighten it. In typical garden quantities this makes matters worse.

Ignoring pore space and focusing only on fertility, then wondering why fertilized plants still struggle in compacted ground.

Working soil when it is wet, which destroys structure and creates compaction that takes years to undo.

Assuming all soil problems are nutrient problems.

## Budget Options

Understanding your soil costs nothing. A hand-feel test - rubbing moist soil between your fingers to judge grittiness, smoothness, and stickiness - gives a usable read on texture for free, and Chapter 3 covers the jar test, which costs the price of a jar.

## Quick Wins

Squeeze a handful of moist soil. If it forms a ball that crumbles when poked, structure is decent. If it forms a sticky ribbon, you have significant clay. If it will not hold together at all, you have significant sand.

Never work or walk on soil when it is wet enough to smear.

## Chapter Checklist

Understand that good soil is roughly 45 percent mineral, 5 percent organic matter, and 50 percent pore space. Know the difference between sand, silt, and clay, and how each behaves. Understand that texture is permanent and structure is improvable. Recognize organic matter as the tool that improves both sandy and clay soils. Stop working soil when it is wet.

## Action Plan

Go squeeze a handful of moist soil from your main growing area and write down what it tells you. That thirty-second observation is the start of knowing your ground.

# Chapter 2: The Living Soil

## What You'll Learn

This chapter covers the biology in your soil - what lives there, what it does for your plants, and what practices help or harm it.

## Why It Matters

Soil is not an inert growing medium that holds plants upright while you feed them. It is a living system, and most of the nutrition your plants receive is delivered by that system rather than directly by you. Growers who understand this stop thinking about feeding plants and start thinking about feeding soil - which is both cheaper and more effective.

## What Lives Down There

A handful of healthy soil contains an enormous population: bacteria, fungi, protozoa, nematodes, arthropods, and earthworms, in a food web where each level eats and is eaten.

Bacteria are the most numerous and drive rapid decomposition of fresh, green, nitrogen-rich material. Fungi decompose tougher, woodier, carbon-rich material and build the thread-like networks that bind soil into aggregates.

Mycorrhizal fungi deserve particular mention. These form partnerships with plant roots, extending far beyond the root system and delivering water and nutrients - especially phosphorus - to the plant in exchange for sugars. A plant with good mycorrhizal associations effectively has a root system many times larger than what you can see. Most vegetable crops form these partnerships. Brassicas and beets are notable exceptions.

Earthworms are the visible indicator species. They consume organic matter, produce nutrient-rich castings, and their burrows create the channels that carry water and air deep into the soil. Counting worms is the cheapest soil health test there is.

## What Biology Does For You

Three main services.

Nutrient cycling. Most soil nutrients are locked in organic matter and mineral forms that plants cannot use. Soil organisms break them down into plant-available forms - a continuous, slow-release fertilization system that runs on its own as long as you feed it.

Structure building. Fungal threads and the sticky substances bacteria produce bind particles into crumbs, creating the aggregate structure and pore space from Chapter 1. Biology literally builds the physical soil.

Disease suppression. A diverse soil community competes with, crowds out, and directly antagonizes many plant pathogens. Biologically active soil simply has fewer disease problems.

## What Helps and What Harms

**Practices That Build or Damage Soil Life**

| Practice | Effect | Why |
|---|---|---|
| Adding organic matter | Strongly helps | Food for the entire food web |
| Mulching | Helps | Moderates temperature and moisture, feeds fungi |
| Keeping living roots in the ground | Helps | Root exudates feed microbes year-round |
| Minimizing tillage | Helps | Preserves fungal networks and aggregates |
| Heavy tillage | Harms | Shreds fungal networks, burns organic matter |
| Compaction | Harms | Destroys pore space; soil life needs oxygen |
| Leaving soil bare | Harms | Erosion, temperature extremes, no root exudates |
| Excess synthetic salts | Can harm | High concentrations stress soil organisms |

## The Shift in Thinking

The practical consequence of all this is a change in approach. Rather than asking "what does this plant need and how do I supply it," you ask "is my soil biology healthy and well-fed, so it can supply what plants need."

That reframing leads to different decisions: adding compost rather than only fertilizer, mulching rather than leaving bare ground, growing cover crops rather than leaving beds empty, and tilling less rather than more. Each of these feeds the system rather than bypassing it, and the effects accumulate year over year.

## Common Mistakes

Treating soil as inert and relying entirely on purchased inputs.

Tilling repeatedly and destroying the fungal networks and structure that took seasons to build.

Leaving beds bare between crops, starving the biology of both food and cover.

Buying mycorrhizal inoculants for soil that would develop its own populations if simply fed and left alone.

## Budget Options

Feeding soil biology costs almost nothing if you compost your own kitchen and garden waste (Book 5 in this series covers it thoroughly) and use leaves, straw, and grass clippings as mulch. The cheapest inputs are often the best ones here.

## Quick Wins

Dig a shovelful of soil in spring and count the earthworms. Fewer than five is a signal; more than ten is a good sign. Repeat annually as a free health metric.

Stop leaving beds bare - plant something, or mulch.

Leave roots in the ground when you clear a crop; cut the tops rather than pulling, and let the roots decompose in place.

## Chapter Checklist

Understand that soil biology delivers most plant nutrition. Recognize mycorrhizal fungi as an extension of your plants' root systems. Know that biology builds structure and suppresses disease. Feed soil rather than only feeding plants. Minimize tillage, keep soil covered, and keep living roots present.

## Action Plan

This week, count earthworms in one shovelful from each growing area and write the numbers down with the date. That is your baseline.

# Chapter 3: Knowing Your Own Soil

## What You'll Learn

This chapter covers assessing your specific soil: the jar test for texture, simple field observations, drainage testing, and reading what your existing plants are telling you.

## Why It Matters

General advice cannot tell you what your ground needs. Two gardens a mile apart can require opposite treatments. These simple assessments cost almost nothing and turn general advice into specific action.

## The Jar Test

The jar test measures your soil's texture - the sand, silt, and clay proportions from Chapter 1.

Fill a straight-sided jar about one-third with soil (removing stones and debris), add water to nearly full, add a small squirt of dish soap to help particles separate, cap it, and shake hard for several minutes. Then set it down and leave it.

The particles settle by size. Sand drops out within about a minute. Silt settles over several hours. Clay can take a day or more, and the water may stay cloudy.

Mark the layers, measure each, and calculate percentages. Worked through: if the settled column is 4 inches total, with 2 inches of sand, 1.5 inches of silt, and 0.5 inches of clay, then you have 50 percent sand, 37.5 percent silt, and 12.5 percent clay - a sandy loam, which drains well and will benefit from organic matter to hold water and nutrients.

**Reading a Four-Inch Settled Column**

| Layer | Measured Depth | Percentage | Settles In |
|---|---|---|---|
| Sand (bottom) | 2.0 inches | 50 percent | About a minute |
| Silt (middle) | 1.5 inches | 37.5 percent | Several hours |
| Clay (top) | 0.5 inches | 12.5 percent | A day or more |
| Total column | 4.0 inches | 100 percent | Read after settling completes |

## The Drainage Test

Drainage is separate from texture and just as important. Dig a hole about a foot deep and a foot across, fill it with water, and let it drain completely. Then fill it again and time how long the second filling takes to drain.

Fast drainage - an hour or less - indicates sandy, possibly droughty soil. Moderate drainage over a few hours is generally good. If water is still standing after many hours, you have a drainage problem, whether from heavy clay, compaction, or a hardpan layer, and no amount of fertilizer will fix the plant problems it causes.

## Reading Your Weeds and Plants

Existing vegetation is free diagnostic information. Take it as a hint rather than a diagnosis, since most weeds grow in a wide range of conditions, but patterns are worth noticing.

Plants that thrive in compacted ground - plantain and chicory among them - suggest compaction. Moss and sedges suggest persistent wetness or poor drainage, and often acidity. Vigorous growth of nitrogen-hungry weeds like lamb's quarters suggests reasonably fertile ground. Widespread stunting and poor color across many species suggests a broader fertility or pH problem.

Your crops tell you things too: consistently yellowing lower leaves suggest nitrogen shortage, purple-tinged foliage can indicate phosphorus problems or cold soil, and forked or stunted root crops often mean compaction or fresh manure.

## Mapping Your Property

Soil is not uniform across a property. Slopes, old building sites, former driveways, drainage patterns, and previous land use all create distinct areas. Walk your ground and sketch a rough map noting where soil looks and feels different, where water collects, where plants grow best and worst, and where there is a history worth knowing about - an old barn site, a filled area, a spot where vehicles parked.

Then test and treat those areas separately, because averaging across genuinely different soils produces recommendations that fit none of them.

## Common Mistakes

Assuming your soil matches the general description of soils in your region.

Testing one spot and applying the results to the whole property.

Diagnosing from weeds alone rather than treating them as a hint.

Ignoring drainage because the soil "looks fine" on the surface.

## Budget Options

Every assessment in this chapter costs essentially nothing: a jar, a shovel, a bucket of water, and attention. Together they tell you more about your ground than most growers ever learn.

## Quick Wins

Run the jar test this weekend; it works while you do other things.

Do the drainage test on your worst-performing bed.

Sketch your property map and mark the good and bad areas.

## Chapter Checklist

Run a jar test and calculate your sand, silt, and clay percentages. Run a drainage test on any suspect area. Read weeds and crop symptoms as hints. Map distinct areas of your property and treat them separately. Record all of it in writing.

## Action Plan

Complete the jar test and the drainage test this week and write the results in a notebook you will keep. Chapters 4 and 5 will turn those observations into a plan.

# Chapter 4: Soil Testing

## What You'll Learn

This chapter covers the laboratory soil test: why it is the best money in gardening, how to take a proper sample, how to read the report, and how often to retest.

## Why It Matters

A soil test is the difference between treating your garden and guessing at it. It costs ten to thirty dollars and typically saves far more than that in amendments you did not need to buy - and it prevents the far more expensive error of adding something that makes matters worse.

## Why Home Kits Are Not Enough

Inexpensive home test kits and pH meters have a place: they give a rough pH reading and can flag an obvious problem. But they are unreliable for nutrients, and they cannot tell you organic matter percentage, cation exchange capacity, or micronutrient status.

A laboratory test through your local extension service gives you actual numbers plus, crucially, recommendations calibrated for your region and your crop. That last part is what makes it actionable. The lab does not just tell you your pH is 5.4; it tells you how much lime to apply, of what type, to reach the target for what you intend to grow.

## Taking a Proper Sample

The test is only as good as the sample, and sampling is where most home gardeners go wrong.

Sample each distinct area separately - the vegetable garden, the orchard, the lawn, and any area with a different history should each get their own test. Do not blend them.

Within an area, take multiple sub-samples - eight to ten spots scattered across it - rather than one scoop from one place. Sample to root depth, generally 6 to 8 inches for garden beds. Use a clean stainless or plastic tool and a clean plastic bucket; galvanized tools and containers can skew zinc readings.

Combine the sub-samples in the bucket, mix thoroughly, and take your submitted sample from that composite. Avoid sampling immediately after adding amendments or fertilizer, and note on the form what you intend to grow, since recommendations depend on it.

**Taking a Sample That Means Something**

| Step | What To Do | Why It Matters |
|---|---|---|
| Divide the property | Separate sample per distinct area | Averaging different soils helps none of them |
| Take sub-samples | 8-10 spots per area | One scoop is not representative |
| Sample depth | 6-8 inches for garden beds | Matches the root zone |
| Use clean tools | Stainless or plastic, clean bucket | Galvanized tools skew zinc readings |
| Mix thoroughly | Composite, then draw the sample | Produces a representative average |
| Note your crop | Write intended use on the form | Recommendations are crop-specific |
| Timing | Fall is ideal | Time to apply slow amendments like lime |

## Reading the Report

A typical report gives you:

pH - the master variable, covered in Chapter 5. Organic matter percentage - your soil-building progress metric; most depleted garden soils run 1 to 2 percent, good garden soil around 5 percent. Phosphorus and potassium levels - usually reported as low, medium, optimum, or high.

Secondary and micronutrients - calcium, magnesium, sulfur, and trace elements. And recommendations - the actionable part.

Note that most tests do not report nitrogen, and this is not an oversight. Nitrogen is highly mobile and changes rapidly with moisture, temperature, and biological activity, so a snapshot measurement has limited meaning. Nitrogen is managed through organic matter, cover crops, and in-season feeding rather than from a test number.

## When and How Often

Test every two to three years under normal conditions, and always in these situations: before starting a new garden, before beginning any significant amendment program, when plants are performing poorly for unclear reasons, and when you want to measure whether your soil-building program is working.

That last use is underrated. Testing organic matter percentage every few years tells you objectively whether your practices are improving the soil or merely maintaining it.

Fall is the ideal time to test, because slow amendments - especially lime, which can take months to change pH - can then be applied with the whole winter to act before spring planting.

## Common Mistakes

Never testing at all and amending by guesswork for years.

Taking one scoop from one spot and treating it as representative.

Blending samples from genuinely different areas.

Testing right after fertilizing, which distorts the results.

Ignoring the recommendations and applying a generic program anyway.

## Budget Options

An extension lab test runs about $10 to $30, which is the highest-return money in this entire series. A home pH kit at $10 to $20 is a reasonable supplement for spot-checking between lab tests, but not a replacement.

## Quick Wins

Order a test today for your main growing area.

Sample in fall so lime has the winter to work.

Keep every test report in one folder; the trend over years is more valuable than any single result.

## Chapter Checklist

Get a laboratory test rather than relying on a home kit. Sample each area separately, with 8 to 10 sub-samples, at 6 to 8 inches deep, using clean tools. Note your intended crop on the form. Understand that nitrogen is generally not reported and why. Retest every two to three years and keep the reports.

## Action Plan

Order a soil test kit from your extension service this week, take a proper composite sample from your main garden, and submit it. Everything in the following chapters becomes specific once that report arrives.

# Chapter 5: pH - The Master Variable

## What You'll Learn

This chapter covers soil pH: what it measures, why it controls nutrient availability, the target ranges for common crops, and how to raise or lower it with real application rates.

## Why It Matters

pH is the single most consequential number on your soil test, because it determines whether the nutrients in your soil are available to plants at all. A garden can be full of phosphorus and still show phosphorus deficiency if the pH is wrong. Correcting pH often solves fertility problems that no amount of fertilizer could fix.

## What pH Measures

pH measures acidity and alkalinity on a scale from 0 to 14, where 7 is neutral, lower is acidic, and higher is alkaline.

The scale is logarithmic, which is the part people miss. Each whole unit represents a tenfold change. Soil at pH 6 is ten times more acidic than soil at pH 7; soil at pH 5 is one hundred times more acidic. This is why a reading of 5.5 versus 6.5 is not a small difference and why pH adjustments must be made deliberately rather than casually.

## Why It Controls Everything

Nutrients dissolve and become plant-available within particular pH ranges. Outside those ranges, they bind into forms roots cannot take up - present in the soil, invisible to the plant.

In strongly acidic soil, phosphorus becomes locked up, calcium and magnesium run short, and aluminum and manganese can reach levels that are toxic to roots. In strongly alkaline soil, iron, manganese, zinc, and again phosphorus become unavailable, which is why alkaline-soil gardeners often see yellowing between leaf veins on susceptible plants.

Soil biology has preferences too - most beneficial bacteria favor near-neutral conditions - so pH affects the whole living system, not just chemistry.

The practical upshot: fix pH first. Adjusting fertility before correcting pH is spending money on nutrients your plants may not be able to reach.

## Target Ranges

**What Different Crops Prefer**

| Crop Group | Target pH | Notes |
|---|---|---|
| Most vegetables | 6.0 to 7.0 | The general garden target |
| Brassicas (cabbage, kale, broccoli) | 6.5 to 7.5 | Higher pH also suppresses clubroot |
| Potatoes | 5.0 to 6.0 | Lower pH reduces scab |
| Blueberries | 4.5 to 5.5 | Strongly acidic; a special case |
| Most fruit trees | 6.0 to 7.0 | Similar to vegetables |
| Asparagus | 6.5 to 7.5 | Tolerates higher pH well |

For a general vegetable garden, 6.0 to 7.0 is the working target and 6.5 is a reasonable bullseye. Note the conflicts in that table: potatoes prefer more acidic conditions than brassicas, which is a genuine argument for grouping crops with similar needs rather than chasing one pH across the whole garden.

Blueberries deserve their own treatment entirely. Their requirement is so far from vegetable range that they are best grown in a dedicated, separately amended bed rather than fighting to acidify a whole garden.

## Raising pH With Lime

Acidic soil is raised with ground limestone. Typical rates, verified against common recommendations:

To raise pH by about one unit in loam, apply roughly 5 pounds of ground limestone per 100 square feet. Sandy soils need less - around 3 pounds per 100 square feet - because they have lower buffering capacity. Clay soils need more, roughly 7 to 8 pounds per 100 square feet, because they resist change.

Worked through for a standard bed: a 4-by-8-foot bed is 32 square feet, so at 5 pounds per 100 square feet you would apply about 1.6 pounds of ground limestone to raise its pH roughly one unit.

Two practical notes. Lime is slow - it can take months to fully act, which is why fall application for spring effect is standard practice. And dolomitic limestone supplies magnesium as well as calcium, making it the better choice when your test shows low magnesium; use calcitic lime when magnesium is already adequate.

## Lowering pH With Sulfur

Alkaline soil is lowered with elemental sulfur, at roughly 1 to 2 pounds per 100 square feet to drop pH about one unit in loam.

Sulfur works through biological action - soil bacteria convert it - so it is slow and temperature-dependent, working little in cold soil. Be conservative: it is far easier to lower pH further later than to correct an overshoot.

If your soil is alkaline because of underlying limestone parent material or hard irrigation water, understand that you are fighting a continuous process. In that situation, growing acid-loving plants in containers or dedicated raised beds with an acidic mix is far more practical than trying to permanently acidify native ground.

## Common Mistakes

Applying lime without a test, "because it can't hurt." It can - overshooting into alkaline territory locks out iron, manganese, and zinc.

Expecting immediate results. Lime acts over months, not days.

Using wood ash casually as a liming agent. It does raise pH, but it is fast-acting, variable in strength, and easy to overdo.

Trying to acidify a whole garden for blueberries instead of giving them a dedicated bed.

Adjusting fertility before fixing pH.

## Budget Options

Ground limestone runs about $8 to $15 per 40- to 50-pound bag, which treats a substantial area - this is inexpensive relative to its effect. Elemental sulfur is similarly modest. The expensive part of pH management is doing it wrong, which is what the soil test prevents.

## Quick Wins

Look up your test's pH result and compare it against the table above for what you actually grow.

If you are liming, do it in fall.

Group acid-preferring and alkaline-tolerant crops into separate beds rather than compromising across the whole garden.

## Chapter Checklist

Understand that pH is logarithmic and controls nutrient availability. Target 6.0 to 7.0 for most vegetables. Fix pH before adjusting fertility. Use roughly 5 pounds of lime per 100 square feet to raise loam one unit, adjusted for texture. Use 1 to 2 pounds of sulfur per 100 square feet to lower one unit. Apply lime in fall and never amend pH without a test.

## Action Plan

When your soil test returns, compare the pH against your crops' targets, calculate the amendment for your actual bed square footage using the rates above, and apply it in fall.

# Chapter 6: Organic Matter - The Engine

## What You'll Learn

This chapter covers organic matter: what it does, realistic targets, how much material it actually takes to move the number, and the sources available to a home grower.

## Why It Matters

If pH is the master variable, organic matter is the master input. It improves structure in every soil type, feeds the biology, holds water and nutrients, buffers pH, and supplies slow-release fertility. Almost every soil improvement in this book runs through it.

## What Organic Matter Does

Structure. Organic matter binds particles into aggregates, which creates pore space. In clay, this opens drainage channels; in sand, it creates water-holding capacity. This is the mechanism behind the apparent contradiction that compost improves opposite soil problems.

Water holding. Organic matter holds many times its own weight in water. Each additional percentage point of organic matter in the top six inches meaningfully increases the water an area can store - commonly cited around 20,000 gallons or more per acre - which translates directly into less watering and better drought resilience.

Nutrient supply and retention. Organic matter both contains nutrients and holds onto nutrients that would otherwise leach away, releasing them slowly as biology breaks it down.

Biology. It is the food supply for the entire soil food web from Chapter 2.

Buffering. It moderates pH swings and cushions the soil against chemical extremes.

## Realistic Targets

**Where You Are and Where to Aim**

| Organic Matter | Assessment | Typical Situation |
|---|---|---|
| Under 2 percent | Depleted | Long-tilled ground, new construction, worked-out beds |
| 2 to 3 percent | Below target | Common in average garden soil |
| 3 to 5 percent | Improving | A working garden with regular additions |
| About 5 percent | Good target | Well-managed garden soil |
| 5 to 8 percent | Excellent | Long-term organic management |
| Above 10 percent | Unusual | Typically peaty soils or very heavy amendment |

The honest note: raising organic matter by even one percentage point is a multi-year project. Soil is heavy, the volumes involved are large, and organic matter is continuously consumed by the biology that depends on it. This is a marathon, and the correct expectation prevents discouragement in year two.

## How Much Compost Does That Take?

The arithmetic is worth working through, because it grounds expectations.

A common annual maintenance application is one-half to one inch of compost spread over the surface.

One inch of compost over 100 square feet requires $100 \times (1 \div 12) = 8.33$ cubic feet, which is 0.31 cubic yards.

Over a standard 4-by-8-foot bed (32 square feet), one inch requires $32 \times (1 \div 12) = 2.67$ cubic feet - roughly two to three of the common bagged sizes.

**How Much Compost You Actually Need**

| Area | One Inch Deep | In Cubic Yards | Notes |
|---|---|---|---|
| 4-by-8 bed (32 sq ft) | 2.67 cubic feet | About 0.10 | Two to three bagged units |
| 100 square feet | 8.33 cubic feet | About 0.31 | Convenient reference figure |
| 200 square feet | 16.67 cubic feet | About 0.62 | Bulk delivery becomes worthwhile |
| 500 square feet | 41.67 cubic feet | About 1.54 | Order in bulk cubic yards |

Note the cost crossover. Bagged compost runs roughly $4 to $8 per cubic foot, while bulk compost runs roughly $30 to $60 per cubic yard delivered - and a cubic yard is 27 cubic feet. Once you need more than about five or six cubic feet, bulk delivery is dramatically cheaper.

## Sources

Your own compost is the best value and the subject of Book 5 in this series. It is free, it recycles waste, and you know exactly what is in it.

Leaves are the most underused resource available to most households - free in autumn, excellent as mulch, and superb composted into leaf mold.

Aged manure is valuable but carries two cautions covered in Chapter 12: it must be well aged or composted before use on food crops, and its source must be known because of persistent herbicide risk.

Purchased compost is convenient and consistent, at a price. Cover crops grow organic matter in place, and Chapter 9 covers them.

## Common Mistakes

Expecting one heavy application to permanently fix soil. Organic matter is consumed continuously and must be replenished.

Adding fresh, undecomposed high-carbon material like sawdust into the soil, which ties up nitrogen (Chapter 7).

Using unknown-source manure or hay and importing a herbicide problem.

Buying bagged compost for large areas when bulk is a fraction of the cost.

Tilling frequently while adding compost, which burns organic matter faster than you add it.

## Budget Options

Home composting and collected autumn leaves cost nothing but effort and can supply most of a home garden's needs. Municipal compost programs often sell finished compost cheaply, though it is worth asking about their feedstock.

## Quick Wins

Collect every leaf you can this autumn; the neighbors will often give you theirs.

Switch from bagged to bulk compost if you are amending more than a couple of beds.

Topdress rather than till in - let biology incorporate it.

## Chapter Checklist

Understand that organic matter improves structure, water holding, nutrition, and biology. Target roughly 5 percent. Accept that raising it a point takes years. Know the quantity math: one inch over 100 square feet is 8.33 cubic feet. Switch to bulk above about five cubic feet. Replenish annually rather than treating it as a one-time fix.

## Action Plan

Measure your growing area, calculate the compost needed for a half-inch to one-inch topdress using the figures above, and price it both bagged and bulk before ordering.

# Chapter 7: Nutrients and Fertilizer

## What You'll Learn

This chapter covers plant nutrients: what N-P-K numbers mean, how to calculate an actual application rate, the difference between organic and synthetic sources, and deficiency symptoms.

## Why It Matters

Fertilizer is where home growers waste the most money and do the most inadvertent harm - over-applying nutrients that were already adequate, under-applying the one that was short, and creating imbalances that cause new problems. A little arithmetic replaces a lot of guessing.

## The Major Nutrients

Nitrogen (N) drives leafy, vegetative growth and is the nutrient most often limiting. It is highly mobile, leaches readily with water, and must be resupplied regularly. Deficiency shows as general yellowing that starts on the oldest, lowest leaves, because the plant moves nitrogen from old growth to new.

Phosphorus (P) supports root development, flowering, and fruiting. It is relatively immobile in soil and its availability is strongly pH-dependent. Deficiency can show as stunting and purplish tints on stems and leaves - though cold soil produces the same symptom temporarily in spring, which fools people every year.

Potassium (K) supports overall vigor, water regulation, disease resistance, and hardiness. Deficiency often shows as scorching or browning along leaf margins, again on older leaves first.

Beyond these are secondary nutrients - calcium, magnesium, sulfur - and micronutrients needed in trace amounts. Calcium deserves a note: blossom end rot in tomatoes and peppers is a calcium availability problem, but it is usually caused by inconsistent watering preventing uptake rather than by a genuine soil calcium shortage. Adding calcium rarely fixes it; consistent moisture usually does.

## Reading the Numbers

Fertilizer labels carry three numbers - for example 10-10-10 - which are the percentages by weight of nitrogen, phosphorus, and potassium.

The arithmetic follows directly. A 50-pound bag of 10-10-10 contains 50 × 0.10 = 5 pounds of nitrogen, plus 5 pounds each of phosphorus and potassium. A 50-pound bag of 5-10-10 contains 50 × 0.05 = 2.5 pounds of nitrogen.

To hit a target rate, divide. A common recommendation is 1 pound of nitrogen per 1,000 square feet. Using 10-10-10, that requires 1 ÷ 0.10 = 10 pounds of product. Using 5-10-10, it requires 1 ÷ 0.05 = 20 pounds of product to deliver the same nitrogen.

**Converting a Nutrient Target Into Product Weight**

| Product | Nitrogen Content | For 1 lb N per 1,000 sq ft | Nitrogen in a 50 lb Bag |
|---|---|---|---|
| 10-10-10 | 10 percent | 10 pounds of product | 5.0 pounds |
| 5-10-10 | 5 percent | 20 pounds of product | 2.5 pounds |
| 20-0-0 | 20 percent | 5 pounds of product | 10.0 pounds |
| Blood meal (about 12-0-0) | About 12 percent | About 8.3 pounds | About 6.0 pounds |

This is the arithmetic that turns a soil test recommendation into an actual amount to measure out, and it is worth doing rather than guessing.

## Organic Versus Synthetic

Synthetic fertilizers deliver nutrients in immediately available form, act fast, are precisely labeled, and are inexpensive per unit of nutrient. Their drawbacks: they leach readily, they feed the plant rather than the soil, they contribute nothing to organic matter, and high concentrations can stress soil biology and burn roots if over-applied.

Organic sources - compost, manure, blood meal, bone meal, kelp, alfalfa meal, feather meal - release slowly as biology breaks them down, feed the soil food web, add organic matter, and are much harder to over-apply harmfully. Their drawbacks: slower to act, more variable in content, and generally more expensive per unit of nutrient.

The approach this series favors is to build fertility through organic matter and biology as the foundation, using targeted amendments for specific deficiencies the soil test identifies. That is not ideology - it is that soil built this way needs progressively less input each year, while soil fed only soluble fertilizer needs the same amount forever.

## The Carbon-to-Nitrogen Trap

One mechanism worth understanding, because it causes a mysterious and preventable problem.

Soil microbes work best at a carbon-to-nitrogen ratio around 24:1 to 30:1. When you incorporate material with a much higher ratio - sawdust at roughly 400:1, straw around 80:1 - the microbes decomposing it pull nitrogen out of the surrounding soil to do the job, temporarily starving your plants.

This is why digging fresh wood chips or sawdust into a bed produces yellow, stunted plants even in otherwise fertile soil. The nitrogen is not gone; it is temporarily immobilized in microbial bodies and will be released later.

The fix is simple: keep high-carbon material on top as mulch rather than mixing it in. Wood chips on the surface do not cause this problem, because the interaction happens only at the thin contact layer. If you must incorporate high-carbon material, add nitrogen alongside it or let it compost first.

## Common Mistakes

Fertilizing without a test and adding what was already sufficient - excess phosphorus in particular is common, persistent, and an environmental problem.

Applying nitrogen to fruiting crops in excess, producing lush foliage and little fruit.

Digging in fresh sawdust or wood chips and causing nitrogen tie-up.

Treating blossom end rot as a calcium deficiency rather than a watering problem.

Diagnosing purple leaves as phosphorus deficiency in spring when the cause is cold soil.

## Budget Options

Compost and cover crops supply a great deal of fertility for very little cash. Where a specific deficiency is identified, a targeted single-nutrient amendment is far cheaper than repeatedly applying a balanced blend of which two-thirds is unnecessary.

## Quick Wins

Calculate the actual product weight for your bed size before applying anything.

Move wood chips from "dug in" to "spread on top."

Check your watering consistency before buying calcium for blossom end rot.

## Chapter Checklist

Know what N, P, and K do and their deficiency symptoms. Convert label percentages into actual nutrient weight. Calculate product weight from your target rate. Understand organic versus synthetic trade-offs. Keep high-carbon materials on the surface to avoid nitrogen tie-up. Fertilize from test results rather than habit.

## Action Plan

Take your soil test recommendation, convert it into pounds of product for your actual square footage using the arithmetic above, and write the figure on the bag so you never have to redo the calculation.

# Chapter 8: Mulch and Keeping Soil Covered

## What You'll Learn

This chapter covers mulching: what covering soil accomplishes, which materials suit which situations, application depths, and the mistakes that turn mulch into a problem.

## Why It Matters

Bare soil is an unnatural state that nature corrects immediately with weeds. Keeping soil covered is one of the simplest, cheapest practices in this book, and it delivers moisture retention, weed suppression, temperature moderation, erosion prevention, and a steady supply of organic matter simultaneously.

## What Mulch Does

Conserves moisture by dramatically reducing evaporation from the surface, which can cut watering substantially in hot weather.

Suppresses weeds by blocking the light weed seeds need to germinate. A good mulch layer eliminates most of the weeding a bare bed requires.

Moderates temperature, keeping soil cooler in summer heat and buffering it against freeze-thaw cycling in winter, which protects roots and reduces frost heaving.

Prevents erosion and crusting by absorbing the impact of rain and irrigation that would otherwise compact and seal the surface.

Feeds the soil as organic mulches break down from the bottom, continuously adding organic matter and feeding the biology - particularly fungi, which favor the woodier materials.

## Choosing a Material

**Matching Mulch to Situation**

| Material | Typical Depth | Best Use | Notes |
| --- | --- | --- | --- |
| Straw | 4 to 6 inches | Vegetable beds | Use straw, not hay (hay carries seed) |
| Shredded leaves | 3 to 4 inches | Anywhere | Free, excellent; shred to prevent matting |
| Wood chips | 2 to 4 inches | Paths, perennials, trees | Keep on surface; superb for paths |
| Grass clippings | 1 to 2 inches | Vegetable beds | Thin layers only, or they mat and sour |
| Compost | 1 to 2 inches | Vegetable beds | Feeds as it mulches; weaker weed suppression |
| Living mulch or cover crop | Not applicable | Off-season beds | See Chapter 9 |

Two material cautions. Use straw rather than hay for vegetable beds - hay is a seed crop and will plant a weed problem you will fight for years. And apply grass clippings thinly, because a thick layer of fresh clippings mats into an anaerobic slab that repels water and smells foul.

## Applying It Correctly

Mulch weed-free, moist soil. Mulching over dry ground locks the dryness in, and mulching over established weeds simply mulches the weeds.

Apply to the depths in the table, understanding that organic mulches settle and thin over the season and need topping up.

And observe the rule that prevents most mulch damage: keep mulch pulled back from plant stems and tree trunks, leaving a few inches of clear space. Mulch piled against stems holds moisture against bark and crowns, invites rot and disease, and gives rodents cover to chew. The "mulch volcano" heaped against tree trunks is a genuinely damaging and extremely common practice.

## Timing

In spring, wait to mulch until the soil has warmed. Mulch applied to cold spring soil insulates the cold in and delays planting - one of the few times mulch works against you.

In summer, mulch heavily to conserve moisture during the hottest stretch.

In autumn, mulch after the ground has cooled to protect against freeze-thaw cycling, and mulch root crops heavily if you intend to overwinter and harvest them in place (Book 16 covers that practice).

## Common Mistakes

Using hay instead of straw and importing years of weed seed.

Piling mulch against stems and trunks.

Mulching cold spring soil and delaying the season.

Laying thick fresh grass clippings that mat and go anaerobic.

Mulching over dry soil or over existing weeds.

Using landscape fabric under mulch in vegetable beds, which impedes the organic matter cycle and eventually fills with roots and weeds anyway.

## Budget Options

The best mulches are frequently free: autumn leaves, grass clippings from your own lawn, and wood chips from tree services who are often glad to dump a load rather than pay disposal fees. Straw is inexpensive by the bale.

## Quick Wins

Call a local tree service and ask to be added to their chip drop list.

Shred and stockpile this autumn's leaves for next season's mulch.

Walk your yard and pull mulch back from every tree trunk and shrub stem.

## Chapter Checklist

Keep soil covered as a standing practice. Match material to situation, using straw rather than hay in vegetable beds. Apply at appropriate depth on moist, weed-free ground. Keep mulch off stems and trunks. Wait for soil to warm before mulching in spring. Top up as organic mulches settle.

## Action Plan

Choose one uncovered bed and mulch it properly this week, then compare its watering needs and weed pressure against an uncovered bed for the rest of the season.

# Chapter 9: Cover Crops

## What You'll Learn

This chapter covers cover cropping: what it accomplishes, which crops suit which purposes, when to sow and terminate, and how to fit it into a home-scale garden.

## Why It Matters

Cover crops are the only soil amendment you grow instead of buy. They build organic matter in place, fix nitrogen from the air, break up compaction with roots, prevent erosion, and feed soil biology - for the price of a few dollars of seed. For a homesteader, that economics is hard to beat.

## What Cover Crops Do

Add organic matter by growing biomass and returning it to the soil, both above ground and - often more importantly - as root mass below.

Fix nitrogen, in the case of legumes. Clovers, vetches, peas, and beans partner with rhizobia bacteria in root nodules that convert atmospheric nitrogen into plant-available forms. That nitrogen becomes available to your following crop as the residue breaks down.

Prevent erosion and leaching by holding soil and roots in place over winter, and by scavenging leftover nutrients that would otherwise wash away - grasses are particularly good at this.

Break compaction with roots. Deep-rooted covers like daikon radish drive channels through compacted layers, then decompose and leave the channels behind as pathways for water and future roots.

Suppress weeds by occupying ground that weeds would otherwise claim.

Feed biology by keeping living roots in the soil, which continuously feed the microbial community through root exudates.

## Choosing a Cover Crop

**Matching Cover Crop to Purpose**

| Cover Crop | Type | Main Benefit | Notes |
|---|---|---|---|
| Winter rye | Grass | Biomass, erosion control | Most cold-hardy; can be hard to terminate |
| Oats | Grass | Biomass, easy termination | Winter-kills in cold climates, leaving a mulch |
| Crimson clover | Legume | Nitrogen fixation | Attractive, good in mild winters |
| Hairy vetch | Legume | Strong nitrogen fixation | Vigorous; terminate before it seeds |
| Field peas | Legume | Nitrogen, fast growth | Good short-window option |
| Buckwheat | Broadleaf | Fast summer cover, pollinators | Frost-tender; seeds fast, terminate promptly |
| Daikon radish | Brassica | Breaks compaction | Deep taproot; winter-kills in cold areas |
| Mixes | Various | Multiple benefits | Grass plus legume is a common pairing |

A useful default for a home garden in a cold climate is a grass plus legume mix - rye with vetch or clover, or oats with peas - which combines biomass and nitrogen fixation in one sowing.

For beginners, oats deserve special mention in cold climates: they grow quickly in autumn, then winter-kill, leaving a mat of dead residue that acts as mulch and is trivially easy to plant into in spring. No termination work at all.

## Timing

The most common home-garden pattern is a fall-sown winter cover: sow after clearing summer crops, let it establish before hard cold, and terminate in spring before planting.

Sow with enough time to establish - generally several weeks before hard frost, and earlier for slower species. Legumes especially need time to establish and begin fixing nitrogen.

Summer covers like buckwheat suit a bed that is between crops for six to eight weeks, and buckwheat's fast growth and pollinator value make it a good gap-filler.

## Terminating

This is where cover cropping most often goes wrong.

Terminate before the crop sets seed. A cover crop allowed to mature and drop seed becomes next year's weed problem, and this is the single most common cover crop mistake. Watch for flowering and act.

Termination methods at home scale: cut and leave the residue as mulch (simple, low-till, effective); crimp or roll to flatten; turn in with a fork or spade; or let winter do it, in the case of winter-killed species like oats and buckwheat.

Then wait before planting. Decomposing residue temporarily ties up nitrogen (the C:N mechanism from Chapter 7), so allow roughly two to three weeks between turning in a cover crop and planting into that bed. Plan the calendar accordingly - this waiting period catches people who terminate the week before they intend to plant.

## Fitting It Into a Small Garden

The honest challenge at home scale is that beds are in production most of the season, and giving one up to a cover crop feels like a loss.

Three practical approaches. Use the off-season, sowing after fall harvest and terminating before spring planting, which costs no production time at all. Rotate one bed per year into a full-season cover crop as a deliberate rest and rebuild. Or undersow, broadcasting a low-growing cover like clover beneath an established tall crop late in its cycle so it is already growing when the crop comes out.

## Common Mistakes

Letting a cover crop go to seed and creating a weed problem.

Sowing too late for establishment before cold.

Planting immediately after turning in residue and hitting nitrogen tie-up.

Choosing winter rye without a plan for terminating it, since it is vigorous and can be difficult at home scale.

Skipping cover crops entirely because beds are "too busy," when off-season sowing costs no production time.

## Budget Options

Cover crop seed runs roughly $3 to $8 per pound, and a pound covers a substantial home garden area. This is the cheapest organic matter you will ever add, since you are producing it on site rather than buying and hauling it.

## Quick Wins

Sow oats this autumn on any bed you are clearing; winter will kill them and leave you a mulch requiring no work.

Sow buckwheat in any bed that will sit empty for six weeks this summer.

Mark your calendar to check cover crops for flowering, so termination never slips.

## Chapter Checklist

Understand what cover crops contribute: organic matter, nitrogen, compaction relief, erosion control, weed suppression, and biology. Match species to purpose, favoring a grass and legume mix. Sow with time to establish. Terminate before seed set. Wait two to three weeks after incorporation before planting. Use the off-season so cover cropping costs no production time.

## Action Plan

Buy a small bag of oats or a rye-vetch mix now and sow the first bed you clear this season. It is the lowest-effort soil improvement available.

# Chapter 10: Compaction, Tillage, and Protecting Structure

## What You'll Learn

This chapter covers protecting the soil structure you build: how compaction happens, what tillage actually costs, the low-till alternatives, and the permanent bed system that solves most of it.

## Why It Matters

Everything in Chapters 6 through 9 builds soil. This chapter is about not destroying it. Growers frequently add compost faithfully while simultaneously tilling and walking on their beds, undoing much of the benefit - and wondering why progress is slow.

## What Compaction Does

Recall from Chapter 1 that roughly half of good soil is pore space. Compaction is the crushing of that space.

The consequences follow directly: water cannot infiltrate, so it runs off or stands; air cannot reach roots or soil organisms, both of which need oxygen; roots cannot penetrate, so plants develop shallow, restricted systems that dry out fast and topple easily; and drainage fails, producing waterlogged conditions in the same soil that will crack hard in drought.

Compaction is caused by pressure on soil - foot traffic, equipment, and especially working or walking on soil when it is wet, when particles slide and pack most easily. This is why the single most protective habit in this chapter is to stay off wet soil entirely.

## The Real Cost of Tillage

Tilling looks productive. It leaves a bed fluffy, level, and weed-free, and the immediate result is genuinely appealing. The costs are less visible and arrive later.

Tillage burns organic matter by injecting oxygen throughout the soil, accelerating decomposition of the very material you have been adding. It shreds fungal networks, including the mycorrhizal associations from Chapter 2, which then take seasons to rebuild. It destroys aggregate structure, replacing built crumb structure with pulverized particles that settle and crust. It brings weed seed to the surface from the seed bank below, which is why heavily tilled gardens often have worse weed pressure over time. And repeated tilling to the same depth creates a hardpan - a compacted layer just below tillage depth that roots and water struggle to cross.

The result is a cycle: till to loosen soil, which degrades structure, which causes compaction, which seems to require more tilling.

## The Low-Till Alternative

The alternative is not "never disturb soil" absolutism - it is minimizing disturbance and letting biology do the incorporating.

The core practices: add amendments on top rather than mixing them in, and let worms and biology carry them down (which they do, reliably). Use a broadfork where loosening is genuinely needed - it lifts and fractures soil without inverting layers or destroying structure. Cut crops at the soil line rather than pulling them, leaving roots in place to decompose into channels and organic matter. And plant into the surface, pulling back mulch to sow or transplant rather than turning the whole bed.

Where tillage is genuinely justified - breaking new ground, incorporating a large amendment into badly degraded soil, addressing severe compaction - do it once, deliberately, and then transition to low-till management rather than making it an annual ritual.

## Permanent Beds: The Simplest Fix

The most effective structural protection available to a home grower is embarrassingly simple: establish permanent beds and permanent paths, and never step on the beds.

This single change concentrates all compaction into the paths, where it does no harm, and leaves growing soil untouched by foot traffic indefinitely. Beds sized so you can reach the center from either side - commonly around four feet wide - make this practical.

**Practices That Preserve What You Build**

| Practice | Effect | Difficulty |
|---|---|---|
| Permanent beds and paths | Confines all compaction to paths | Easy, one-time layout |
| Never work wet soil | Prevents the worst compaction | Free, requires patience |
| Topdress instead of tilling | Preserves structure and fungal networks | Easy |
| Broadfork instead of rototill | Loosens without inverting | Moderate effort |
| Cut crops, leave roots | Adds organic matter and channels | Easy |
| Keep soil mulched or covered | Prevents crusting and erosion | Easy |
| Deep-rooted cover crops | Biological decompaction over time | Easy, slow |

## Common Mistakes

Working or walking on soil while it is wet.

Rototilling annually out of habit.

Adding compost faithfully while tilling it away just as fast.

Walking through the middle of growing beds for convenience.

Expecting a single deep tillage to permanently solve compaction; without changed practices it returns.

## Budget Options

Permanent beds and staying off wet soil cost nothing. A broadfork runs roughly $150 to $250 and lasts a lifetime, but it is an upgrade rather than a requirement - a garden fork used carefully accomplishes much of the same work.

## Quick Wins

Lay out permanent paths this week, even if only with mulch or boards, and commit to staying off the beds.

Cancel the annual tilling and topdress instead for one season, then compare.

Cut your next spent crop at the soil line rather than pulling it.

## Chapter Checklist

Understand compaction as the destruction of pore space. Know tillage's real costs: burned organic matter, shredded fungal networks, destroyed structure, weed seed exposure, hardpan. Adopt low-till practices: topdress, broadfork, cut rather than pull. Establish permanent beds and paths. Never work wet soil.

## Action Plan

Mark out permanent beds and paths in your garden this week and commit to never walking on the growing areas again. It is the single highest-return structural decision available.

# Chapter 11: Diagnosing Soil Problems

## What You'll Learn

This chapter covers reading symptoms: distinguishing nutrient problems from pH problems from physical problems, common look-alike confusions, and the specific contamination risks worth testing for.

## Why It Matters

Most soil problems get misdiagnosed, and a misdiagnosis costs money and a season. The most common error is treating a physical or pH problem as a nutrient shortage - adding fertilizer to soil that is compacted, waterlogged, or locked up at the wrong pH.

## The Diagnostic Order

Work through possible causes in this order, because the earlier ones are more common and cheaper to check.

First, physical. Is the soil compacted, waterlogged, or too dry? Is drainage adequate (Chapter 3)? Physical problems produce symptoms that look exactly like nutrient deficiencies, because roots that cannot function cannot take up nutrients regardless of how much is present.

Second, pH. Is pH within range for the crop? Wrong pH locks out nutrients that are physically present. Check this before buying any fertilizer.

Third, nutrients. Only after ruling out physical problems and pH does a genuine nutrient deficiency become the likely explanation - and a soil test tells you which.

Fourth, other causes. Disease, pests, herbicide damage, root damage, weather stress, and variety characteristics all produce symptoms that mimic soil problems.

## Reading Symptoms

**What Common Symptoms Usually Mean**

| Symptom | Likely Causes | Check First |
|---|---|---|
| Yellowing on oldest, lowest leaves | Nitrogen shortage | Nitrogen; also waterlogging |
| Yellowing between veins on new leaves | Iron or manganese unavailability | pH too high |
| Purple tint on leaves and stems | Phosphorus, or simply cold soil | Soil temperature first, in spring |
| Browning or scorched leaf margins | Potassium shortage; salt injury | Potassium; also over-fertilization |
| Stunting across many species | pH, compaction, or contamination | pH and drainage |
| Wilting despite moist soil | Root damage, waterlogging, disease | Drainage and roots |
| Forked or stunted root crops | Compaction, stones, fresh manure | Physical soil condition |
| Blossom end rot | Calcium uptake failure | Watering consistency, not soil calcium |

## Look-Alikes Worth Knowing

Four confusions cause most misdiagnoses.

Cold soil mimics phosphorus deficiency. Purple-tinted spring seedlings are usually reacting to cold, not shortage, and green up as soil warms. Wait before amending.

Waterlogging mimics nitrogen deficiency. Roots in saturated soil cannot take up nitrogen, so plants yellow exactly as if the soil were poor. Adding nitrogen to a drainage problem accomplishes nothing.

Inconsistent watering mimics calcium deficiency. Blossom end rot is the classic case, discussed in Chapter 7.

Herbicide damage mimics disease. Twisted, cupped, distorted new growth - particularly on tomatoes, beans, and other sensitive plants - often indicates herbicide exposure rather than pathogens, and Chapter 12 covers the manure-and-compost route by which it commonly arrives.

## Contamination: What to Actually Worry About

Two contamination risks deserve genuine attention rather than vague anxiety.

Lead is the significant one for home food growers. It accumulates in soil near old painted structures (from decades of paint chips and dust, as this series' exterior volume explains), along old roadways from leaded gasoline, and on sites with industrial history. Children and root crops are the primary concerns.

If you garden within a few feet of an old house foundation, on an urban lot, or on ground with unknown history, have the soil tested for lead - most extension services offer it as an add-on for a modest fee. If levels are elevated, the practical responses are growing in raised beds with clean imported soil, maintaining a high organic matter content and near-neutral pH (both of which reduce lead availability), washing produce thoroughly, peeling root crops, and keeping children out of bare contaminated soil.

Persistent herbicides are the other, and they arrive through imported material. Certain broadleaf herbicides used on pasture and hay fields - the aminopyralid family among them - survive grazing, composting, and manure aging, and remain active in the resulting compost or manure for a long time. Applied to a garden, they severely damage or kill tomatoes, beans, peas, and other sensitive crops, sometimes for multiple seasons.

The protection is simple: know the source of any manure, hay, straw, or bulk compost you bring in, and ask directly whether the pasture or field was treated. When in doubt, run a bioassay - plant beans or peas in a pot of the material and watch for twisted, cupped growth before spreading it across a garden.

## Common Mistakes

Fertilizing to fix a compaction or drainage problem.

Amending pH or nutrients based on symptoms without a test.

Diagnosing spring purple leaves as phosphorus deficiency.

Bringing in unknown-source manure or compost.

Growing food in soil with an unknown industrial or old-structure history without testing for lead.

## Budget Options

The diagnostic sequence is free: check physical conditions, then pH, then nutrients. A lead test is a modest add-on to a standard soil test. A herbicide bioassay costs the price of a few bean seeds and a pot.

## Quick Wins

Before buying any amendment, check drainage and compaction on the affected bed.

Run a bean bioassay on any new bulk compost or manure before spreading it.

If your house predates 1978 and you garden near the foundation, add the lead test to your next soil test.

## Chapter Checklist

Diagnose in order: physical, then pH, then nutrients, then other causes. Know the common look-alikes. Test for lead if the site history warrants it. Know the source of imported manure and compost and bioassay when uncertain. Stop treating symptoms without identifying causes.

## Action Plan

Pick your worst-performing bed and work the diagnostic order on it deliberately - drainage, compaction, pH, then nutrients - before spending another dollar on amendments.

# Chapter 12: Your Multi-Year Soil Plan

## What You'll Learn

This final chapter assembles everything into a realistic long-term program: what to do in year one, what to expect by year three, an annual rhythm, and how to measure progress.

## Why It Matters

Soil building fails most often from impatience - a grower does everything right for one season, sees modest change, and concludes it did not work. Understanding the actual timeline keeps you going long enough for the compounding to show, which is when the results become dramatic.

## The Honest Timeline

Year one is diagnosis and correction. Test, fix pH, address any drainage or compaction problem, establish permanent beds, and begin adding organic matter. Results are modest and mostly invisible. This is the year that requires faith.

Year two brings visible change. Soil works more easily, holds moisture better, and plants perform noticeably more consistently. Earthworm counts rise.

Year three is where the compounding becomes obvious. Structure is genuinely improved, watering needs drop, fertility needs decline, and problems that used to be routine simply stop appearing.

Year five and beyond is maintenance rather than construction. Annual topdressing and cover cropping hold the gains, and the garden becomes materially less work than it was in year one.

That trajectory - hard at first, then progressively easier - is the opposite of how most garden effort works, and it is why soil is the best place to invest.

## The Annual Rhythm

**A Repeating Annual Program**

| Season | Primary Soil Work | Notes |
|---|---|---|
| Late winter | Plan from test results; order amendments | Review last season's records |
| Early spring | Topdress compost; broadfork only if needed | Do not work wet soil |
| Spring | Terminate cover crops, wait 2-3 weeks, plant | Mulch after soil warms |
| Summer | Maintain mulch; side-dress heavy feeders | Keep soil covered |
| Late summer | Sow cover crops as beds clear | Oats and peas are easy starters |
| Autumn | Soil test; apply lime if needed; collect leaves | Fall lime has winter to act |
| Winter | Beds under cover crop or mulch | Never leave soil bare |

## The Priority Order

If resources are limited, do things in this order, because each step makes the next more effective.

First, test. Ten to thirty dollars replaces years of guessing.

Second, fix pH. Everything else depends on it, and it is cheap.

Third, fix physical problems. Drainage and compaction defeat all fertility work.

Fourth, establish permanent beds and stop tilling. Free, and it protects everything you build.

Fifth, add organic matter annually. The engine of long-term improvement.

Sixth, keep soil covered with mulch and cover crops year-round.

Seventh, correct specific nutrient deficiencies the test identifies.

Notice that the first four steps cost very little money, and that fertilizer - where most growers start - is last.

## Measuring Progress

Keep records, because soil change is slow enough that memory is unreliable.

Track organic matter percentage from your periodic soil tests; it is the single best objective progress metric. Track earthworm counts from a standard shovelful each spring. Note how the soil works - whether it crumbles or clods, how easily a fork goes in. Record watering frequency, which should decline measurably as organic matter rises. And note problems that stop appearing, which is often the first sign a grower notices.

## Common Mistakes

Quitting after one season because change was modest.

Starting with fertilizer instead of a test.

Building soil in one bed while continuing to till and walk on the rest.

Failing to keep records, then being unable to tell whether anything worked.

Chasing a single miracle amendment instead of running a consistent program.

## Budget Options

The entire priority list through step six is achievable on a very small budget: a soil test, a bag of lime if needed, home compost, collected leaves, and a few dollars of cover crop seed. The expensive approach - buying fertility every year - is what this program gradually replaces.

## Quick Wins

Write your year-one plan on one page: test, pH, drainage, permanent beds, compost, cover crop.

Start a soil notebook with today's earthworm count and your current test results.

Order the test if you have not.

## Chapter Checklist

Accept the multi-year timeline and expect modest year-one results. Follow the annual rhythm. Work the priority order: test, pH, physical, permanent beds, organic matter, cover, then nutrients. Track organic matter, earthworms, workability, and watering frequency. Keep written records.

## Action Plan

Write your one-page soil plan this week, start the notebook, and commit to running the program for three full seasons before judging it.

# Conclusion: The Ground Beneath Everything

There is a reason this book sits where it does in the series rather than at the front. Most people arrive at soil the long way around - after the disappointing tomatoes, the bolted lettuce, the season that just did not work - and only then start looking down instead of out.

But once you do look down, the logic reorders itself. The seed catalog matters less. The fertilizer aisle matters less. What matters is a slow, patient, compounding practice: test the ground, correct what is genuinely wrong, feed the biology, keep the surface covered, stay off it when it is wet, and repeat that until the soil does most of the work for you.

That is the quiet trade this book offers. Effort front-loaded into the ground, returned with interest every season after, in the form of plants that simply grow - with less watering, less feeding, and fewer of the problems that used to define the gardening year.

Test your soil. Fix your pH. Stop tilling. Feed it, cover it, and give it three years.

Everything else in this series grows out of that ground.

# About the Author

Hudson Briggs writes practical, plainspoken guides for people who want to do more for themselves - growing, building, preserving, and repairing - without pretending that everything is easy or that shortcuts are free. The aim in every book is the same: honest numbers, real costs, clear steps, and a straight answer about what a beginner can genuinely expect.

The Homestead Starter Series

Book 1 - The Beginner's Guide to Homesteading

Book 2 - Raised Bed Gardening

Book 3 - Backyard Chickens

Book 4 - Food Preservation

Book 5 - Composting Made Simple

Book 6 - Homestead Projects

Book 7 - Small Livestock for Beginners

Book 8 - Homestead Budgeting

Book 9 - Water Systems

Book 10 - Seed Saving and Propagation

Book 11 - Meat Processing and Curing

Book 12 - Off-Grid Energy

Book 13 - Selling from the Homestead

Book 14 - The Herb Garden

Book 15 - The Backyard Orchard

Book 16 - Season Extension and the Home Greenhouse

Book 17 - Soil and Fertility